MOBILE SOLAR

POWER SYSTEMS

For

VANS and RVs

MOBILE SOLAR POWER SYSTEMS

For

VANS and RVs

Power Up to Go Off Grid.

Joe Kandell

Table of Contents

Introduction

This book has been written for the person who has never installed a solar unit on their RV, van or vehicle they own and want to live off the grid.

Prices are coming down on solar devices, and there are so many options available for you to choose from now that you can start on the small and gradually build up to large. You will be able to live big or live frugally - it is all up to you. And this book will tell you how to do just that.

You will not find any shortcuts in this book because it is crucial that when you finish your project that you will be proud of it and that it will last for years to come.

I will be honest with you throughout the book and tell you that the money you spend on your solar system, in the beginning, will be recouped in the years to come. As you read, you will see the pros and cons of the different type of components available so that you will not be left with any surprises.

You will not have to be an expert electrician to take on this project. Just follow the directions, and you will be prepared to handle the tasks at hand. You will probably enjoy this project as you work toward finally being 'off the grid.'

I wish you luck on your project because I know you will be successful using the strategies outlined in this book.

Chapter One

Should We or Should We Not Live Off the Grid in Our RV or Van?

Here we are, trying to decide if we want to take that big leap. Sounds a bit daunting, doesn't it? You may say to yourself, "But I do not know anything about electricity and wiring!" Don't worry; neither does 85% of the civilized world. We may know that if we turn on a light switch without power that the bulb is no longer any good if a fuse has blown, or maybe, just maybe we need to throw a breaker.

However, there is hope for all of us! As we start to take this journey one step at a time, we will work through this process together. By the time you are done, I have a feeling this will seem like a piece of cake.

At first, when you start thinking about all the work it will take to "go to the dark side," and live "off the grid," it could sound a little frightening. All this means for you is that you will be making your own power for the most part, and own some great solar panels on your mobile unit.

Don't think of it as 'all glory' immediately, because there will be, as with most things, an initial start-up cost and some labor involved. It will surprise you at how much you can save over the years.

It all depends on if you are living in your RV or van full time, or just using them for vacationing and weekend getaways. I like the idea of just being "off the grid" and not having to depend on anyone for my electrical needs. Then you won't be at anyone's mercy when they decide to double their rates.

If you are living in a large RV and considering that solar panels have a lifespan of about twenty years, you might be looking at saving as much as $20,000 in that length of time.

By taking your energy mobile, you can live and travel anywhere you desire, with a different view of the sky every night.

With global warming, our climate could see a significant change during the next century. If you can stay "off that grid," it will help to lower the emission rates around the entire world. YOU can make a significant impact by leaving the grid. For example, over a period of 30 years, if you would use solar energy, you and you alone, could make the difference of 178 tons *less* of carbon dioxide. Wouldn't that leave a fantastic mark on earth for the next generation?

Going solar can be expensive. It has come down significantly to from what it was years ago, but be prepared to spend your most money upfront. There are a wide variety of price ranges which we will go over later in the book, but you will see that this money will be recouped over time.

Think about the mobile solar power unit and the freedom it can give you. If you get tired of living in one area of the country, it is no problem to pick up and move. With jobs becoming more mobile all the time, so can you.

You won't have to leave a forwarding address – you don't have to call the electric company and tell them to unhook you. When you arrive at your next stop, you don't have to go to the local electric company, pay a deposit, and set up a new account at the new location. Why? Because you are living 'off the grid' and making your own energy, so you don't need to buy theirs anymore

Another item that deserves your thought is that there are specific locations in the world that are just better for solar power than others. If you are thinking about setting up in a cloudy part of the world, then you better know that making energy is going to be like watching a snail crawl. Not good.

If you are looking for tax breaks, only certain states give them, and it can change from year to year. Then, there are the lucky few who live in places where it is cheaper to stay on the grid than to go off the grid. I can't imagine where that would be compared to where I live, but maybe there is such a faraway electric land I have never heard about.

Going solar will also involve a LOT of commitment from you. I want to make this clear in the very beginning. By no means, am saying that it will take up all your time. It will not. But you must be dedicated to what you are doing to do this right. It is somewhat complicated, and you will need to do your research and seek information from fellow solar energy users.

Your solar energy system will require regular maintenance. If you live in a dry climate, you may have to wash your panels more frequently. If you live in an area around a lot of farmland, I promise you that you will be washing them a lot. It will be essential to get an annual maintenance checkup so that you can be sure your panels are operating at 100%.

If you are genuinely dedicated to making this work, you will need to take a sharp look at your current living situation, lifestyle, habits, and probably make some tough decisions on how you will conserve energy in your life.

Living "off the grid" full time requires a lifestyle change, a lot of dedication to the project, extra effort, work and if there are children, it will take family buy-in for the project to succeed.

Here is a list of pros and cons to living off the grid.

PROS	CONS
Technology is still improving	Requires space for set-up
Solar is low maintenance	Intermittent at times if cloudy
Financial incentives in some states/and federal	Storage for energy is expensive in beginning
System runs quietly	Is slightly associated with pollution
Reduction of your electrical costs	Constructed of exotic materials
Solar is readily available	Expensive
Great for the environment	
It is sustainable	
The sun is abundant	
It is renewable without waste	

Chapter Two

Lead Acid Batteries vs. Lithium Batteries

You need to look at the different types of batteries there are to store your energy in for your solar power system. Don't make snap judgments about what you want until you have thought about your investment and what will work best for you over the long haul.

There are two types of Lead Acid Batteries. There are the Flooded Lead Acid and the VRLA (valve regulated lead-acid). You will find that the Lead Acid Batteries are about $65 each, the VRLA Batteries are about $120 each, and your Lithium-ion (LiNCM) are about $600 each.

- We already know that a lead battery is just about as heavy as two concrete blocks tied together. Nearly three times heavier than a lithium battery.

- If we are to compare regarding efficiency; we will have to say that the lead-acid is just nothing but inefficient in charging and discharging. It means losing amps when you charge your battery.

- And, you will notice a fast voltage drop when it is discharging, and that will decrease the overall capacity of that battery. It is something that just doesn't occur with Lithium-ion batteries.

- If you do choose a lead-acid battery (made of lead plates and lead oxide that sits in a solution of electrolyte) remember that it is made up of 35% sulfuric acid and 65% water. You will find that the specific gravity of the solution gets higher as your battery charges and lower as it discharges. When the battery starts discharging, the sulfur begins moving away from the

solution and back toward the plates. Then as the battery starts charging again, the sulfuric acid will move out into the solution causing the specific gravity to rise again.

- One thing that you will certainly notice is that a lead-based battery will never discharge over 80%. You will find that most of your manufacturers will tell you not to discharge them lower than 50%. When it comes to the Lithium Battery, you will almost always get 100% of discharge.

- We all know we want our batteries to live as long as possible. Most of the lead batteries will only have a lifetime of 300 cycles, but your lithium battery will probably live as long as 700 to 800 cycles. Cycle life in batteries is exceptionally dependent on the discharge level. It takes a toll on lead-acid batteries but barely affects lithium-ion batteries.

- If you are watching your voltage levels, you will notice that with lead batteries there is a continual drop of voltage and with the lithium-ion, the voltage levels stay constant. If you can maintain your voltage, it will be even more efficient.

- As mentioned earlier, the lead batteries are the cheapest, but with lithium-ion, once you factor in the life of the battery and the performance it is the overall winner.

- There is a newer lithium battery on the market called LiFe battery. LiFe is short for lithium – iron – phosphate. This battery will hit 2500 charge cycles. WOW! The battery has immunity to the "charge memory." They cost you more money up front, but they will still be working when you have thrown away eight of the lead batteries.

They are much safer and not like lead or the standard lithium that can catch fire or leak out gas. These mobile units are *so easy* to install.

The electrolyte specific gravity depends on the 65% to 35% ratio for the precise chemical reaction to come in to play. The ratio can be

affected by temperature and the amount of sulfuric acid of the solution in the battery.

You will notice that when the temperature drops, the electrolytes in the solution of the battery will contract causing the liquid to be denser due to the number of electrolytes that can now be present in the liquid. The denser the solution, the higher the specific gravity.

When the temperature rises, the electrolytes expand, and the solution cannot store as many, so it will not be as dense, therefore lowering the number of the specific gravity.

For a battery, the best way to test its charge is by checking its specific gravity. It is true that while discharging, specific gravity will continue to decrease linearly as the ampere-hours are released. Then, as the battery starts to recharge, the specific gravity will increase again.

Checking the Battery Charge by Using the Hydrometer

The hydrometer is a method to test and measure for the specific gravity of the solution in each cell of the battery. It measures the weight or density of any liquid compared to the density of the same amount of water. When a specific gravity is 1.265 at 80 degrees Fahrenheit, you will see that a lead acid battery cell is at full charge.

If you need to adjust for temperature variances, take the specific gravity reading and correct for the temperature for every ten degrees above 80 degrees Fahrenheit and add .004 and then for every 10 degrees that falls below 80 degrees Fahrenheit subtract .004.

Here are some tips to remember for properly using the hydrometer.

- Always check more than one cell when you are testing for the state of charge. For me, I always check every cell in the battery in the event there is a bad cell.
- If you have just watered the battery, do not expect an accurate reading.

- For the reading to be accurate, the electrolytes should be at room temperature.

- When you are working on the battery, wear protective eyewear.

- Electrolyte residue that might stick to the hydrometer is corrosive, do not forget that, so rinse it off and store it in an acid-resistant container.

There are two other ways to check the charge of your batteries. One is by an amp-hour meter. It only gives you the state of charge and is not as reliable as measuring by specific gravity.

The other way is by measuring the voltage of the entire battery bank to determine the state of charge which is not as reliable as measuring by specific gravity.

No matter what you do with your batteries, it seems one of the hardest parts of this entire project is trying to figure out where to put them whether you have two or four. Where will they fit, where will they have to go, and where will the wires have to run so they will have the shortest path (under eight feet).

It is essential to equalize the batteries that you are using. The batteries you use for your solar system is sometimes made of several batteries, and each battery has one or many cells. After the batteries are charged, you may find that the specific gravity is different in the cells of the battery bank. By equalizing them, you can bring every cell up to a full charge. If you balance the cells, it reduces the sulfation and the stratification taking place in each cell.

Most battery companies think that you should equalize your batteries every six months. Some recommend using a hydrometer to test each cell for its specific gravity after you have done a full charge. Doing this can show if any of them show a significantly lower specific gravity than the others, in which case it is best to do an equalization.

The charge controller that is fed by solar power usually does the equalization. However, at times, if the inverter has a charger your generator feeds it. The charger might let you enter a voltage for the equalization and the length of time in doing it. The charger will begin charging the battery bank until the time your voltage is reached and will stay there for the set time.

If you are using flooded lead-acid batteries in your RV, they contain water and sulfuric acid. When a battery undergoes charging, or a load is being taken off the batteries, some of the water will break down into a gaseous form that can escape from the batteries, allowing a gradual loss of water.

Now, a sealed battery does not have this problem. If you have non-sealed batteries, you can purchase hydro caps that will replace the caps that came with the batteries on each of the cells of the battery. The hydro caps will capture the hydrogen and oxygen, so they cannot escape, and turn them back into water that falls back down into the battery. Even with hydro caps, you will have some steam escape, and thereby some water loss.

If your battery is not sealed, then you are going to have to check the fluid level in your batteries and top them off as it is needed. You NEVER need to add sulfuric acid. Each battery has more than one cell usually, and you will need to take the cap off to pour in the water. Always use distilled water for this task. Clean the battery cell cap before removing it so that there is no debris to fall inside your cell.

When you are filling, watch how much water you are putting in; you don't want the water to be below the lead plates or above the lead plates. Top the battery off with like I said, distilled water ONLY. You can usually buy this at a grocery or drug store.

How often you will need to do this will depend on how much you are using your system. If you have a new system, check it at least once a week. The more you get used to it, you will soon learn how often you need to perform this maintenance.

You will find that the gases that escape from the battery come out through the cap and settle down on the top of the battery. When it settles there, it can become electrically conductive and can create problems and act like a small load always turned on.

So, it needs to be cleaned off. Take baking soda and pour a thin layer on the battery where the film has settled. You will be able to see the baking soda as it neutralizes the acid because it will fizz. After the fizzing stops, you can take a damp rag and wipe it off. Then you can rinse the rag off in the sink.

Just remember this important tip. Wherever you store your batteries, it must be _vented_.

The following is a great link that I feel will help you a lot. It gives you the actual tool to use with either a 12 or a 24-volt DC system using a DC to an AC through an inverter.

https://www.batterystuff.com/kb/tools/solar-calculator.html

Chapter Three

A Brief Lesson in Electricity

You will find two types of equipment that you see in the RV that use electricity - AC, and DC. The DC equipment will likely run off the battery, including things like your water pump, the lights, your vent fans, and about anything else you could plug into a 12V plug-in.

The AC items are your microwave and air conditioner. You will get power for them by putting a big plug into a plugin at the RV Campsite or by running a generator. But that generator is very noisy and drinks about one half to a gallon of gas an hour from your main tank. You hate running it because it is expensive to run and so loud that you can't carry on a conversation.

A significant term to understand when dealing with your batteries is Watt-hours. Let's take Watt-hours and think of them like they were a bucket of electric, and let's say Watts is a hole in our bucket.

My batteries can store about 2400 watt-hours. So, I could burn a 24-watt bulb for 100 hours. Or, if I wanted I could watch a 200 watt TV for 12 hours if I were binge-watching Grey's Anatomy.

Amps are nothing but Watts that are divided by volts. On a 12 volt system like the kind you would have in an RV DC, 1 amp would be equal to 12w.

Your most straightforward way to work with all of this is to multiply your amps by volts and then work in watts.

The only power you can store is DC in a battery. You can't save AC power. If you wanted to turn on one of your appliances that were AC, you would have to get an inverter so that it would convert your DC to AC.

Herein lies the problem - devices that are supposed to work off AC power use a lot of watts, since there is a lot of AC power.

Calculating Energy Usage

Here is a list of watt usage:

- When you are charging a laptop with empty battery – 60w
- Using laptop when battery is not charging – 20w
- Using a light bulb – 20w
- Water Pump when actively being used – 50w
- Vent fan (it depends on the speed) 12-35w

Just for kicks, let's see how much one person could use of electricity in one day. Myself, I am terrible with electricity, and my bill shows it. In this example, I applaud this person that is extremely frugal in their energy usage.

Say, for instance; there is one light on from 10 pm to 3 am.

Five hours x 20w. = 100 watts

With a laptop on for that same timeframe, there are another 100 watts

The fan will run for about 11 hours a day on low and 13 hours a day on medium.

130 + 285 = 415 watts

How about an hour for using the water pump for dishes and the bathroom? 25watts

Plug in the laptop for about 6 hours. 120 watts

So, we add it all up, and we have used about 760 watts a day. So how do I get that power? Now seriously, the loud generator could be started up and ran it for a while when you needed power. Or, you could get out and drive a couple of hundred miles, and that would charge the batteries up from the inverter on the RV. Or, better yet, we could go with a complete solar system for our RV.

Let's look at a few more examples:

A radio uses 10w, and you have it on for 5 hours a day = 50w each day

A water pump uses 20w and is on for 20 minutes a day = 6.66w each day

Your main light in the RV living area is 30w, and it is on 3 hours a day = 90w each day

Total used = 146.66w each day

Before you do anything else, swap all your lights out for *LED lights*. They will help you tremendously as they *use 80% less energy* and give out even better light.

You will need to look at how much your solar panels will need to generate for you each day. Calculating the energy your panels supply to your batteries is done with this equation:

Watts (of your solar panel) x hours exposed to sunlight = numbers of watts sent to battery

Say we are going to figure this up for being out in the late springtime and stay to early Autumn to figure our equation.

Watts required / time of year hours of sunshine = panel size -> 146.66/6 = 24.4W panel

Since they do not make a panel this size, we would go up to the next size panel which would be a 30w or even a 40w solar panel.

Now we need to know how much energy our panel can generate. Here is our formula for that:

(the rated power of your solar collector(panel) X (direct sunlight in number of hours) X (your best guess of time it will be receiving direct sunlight with no clouds) X (the efficiency of your charge controller)

Let's fill this example in and see what we come up with:

(80 W) X (5 hours of sun) X (70% direct sun) X (85% efficiency of your charge controller) = 238 WH (watt-hours)

Here comes the solar miracle. Depending on where you are on this great planet of course, we know that in this instance we will need 760 watts a day. Most days where this person lives, there will be about 5 hours of sunshine. We also understand that a 1-watt panel will only make me 5 watts a day. That is not going to be enough even to run an electric toothbrush. Maybe a 156-watt board would work, but nope - that is just 780 watts a day. We do not want to cut it that short. We need some excess power. We might need more electricity to be prepared for cloudy days. It would be a good idea to purchase a 200-watt panel and if you should want to later, add another board.

You can't just take that solar panel and stick it up with command strips to the back or top of your RV and run a wire to your battery. Even a 12v board can put out 17v of sunlight, and that is overkill for a battery. Here is where the charge controller comes in to regulate the charge.

Will you need to be sure you have the right batteries to store all this energy in now? Batteries are rated in what they call amp-hours, and that is mainly watt-hours that are divided by 12.

Hopefully with the 200-watt solar panel you will be able to generate 1000 watt-hours every day. It means that *you will need about 2000 watt-hours of capacity in the batteries. Divide that by 12* to make up your amp-hours, and it will come to 166Ah. There you go, *"now you have the 116.6 amp hours."*! (See configuration below with sun exposure of 7 hours on this day.) Most RV's have two batteries, and a lot of people use their RV's as their primary form of transportation, so as they drive they are charging their batteries.

Amp hours = Watt hours/12

now let's add our numbers

116.6 amp hours = 1400/12

In this process, inverters play a crucial role. They do a simple job for you. They take electricity from 12v DC and convert it into 110v AC. They pull the electricity from the batteries and make it usable for your coffee maker, computer, and to charge your cell phone.

There are TWO types of power inverters: a modified sine wave inverter and a pure sine wave inverter. The modified sine wave inverter can be as simple as a 75-watt modified that is very small and will plug right into your cigarette lighter. It can run your laptop with ease. Most of your items that have motors and some of your electronics will need this pure sine wave inverter.

The pure sine wave inverter can produce power that is almost identical or even better than what you get from the power company. When you purchase your inverter, make sure the wattage has a rating high enough, so it will be able to power all the things you want to plug into it. It is always a good idea to buy one larger than what you think you will need.

Chapter Four

PWM and MPPT Charge Controllers

PWM stands for *pulse width modulating*. It is a charge controller that uses a set of sophisticated algorithms to figure out how much charge is going into a battery and slowly start it to taper off the charging when the battery reaches full.

The PWM solar chargers utilize a technology that is similar to other modern battery chargers of high quality. When you have a battery voltage that reaches a specific amount, the PWM formula will slowly reduce charging current to avoid gassing and heating the battery, but charging will continue by returning to the fullest volume of energy back to your battery in a short amount of time. It results in rapid recharging, one or more healthy batteries to full capacity, and higher charging efficiency.

You can use a PWM controller with panels of higher voltage (say 20 volts or more) just so long as the voltage will not exceed the maxed out input voltage of that controller. Remember though, only a part of the power of the high voltage that comes from the solar array will actually pass through the PWM controller and go on to your battery bank. If you have high voltage panels, you probably need an MPPT controller instead.

PWM charges your batteries in three stages.

- During the **BULK** phase of the charging process, the voltage will rise slowly to around 14.4 volts while your batteries are drawing maximum current. When this level is reached, your next stage will start.

- **ABSORPTION** phase where the voltage is kept at the bulk level for a specific amount of time while current will taper off gradually when the batteries charge up.

- Then the **FLOAT** phase sets in after the absorption time is over, and the voltage is decreased to float level (13.4 volts), and the batteries start to draw a small amount of maintenance current before the next cycle.

However, this does have a drawback. For the PWM controller, you must size the system so that it will match what the voltage is of your battery bank, and for the most part, PWM controllers are limited to about 60 amps as their max.

MPPT stands for *multiple power point tracking*. It is another type of charge controller that factors in variables that can change the output that happens due to the weather conditions. It has the capability to match up automatically the voltage being produced to the battery voltage to max out the charging efficiency.

Here is another way to put it. MPPT looks at what your panels are putting out and then compares it to your battery voltage. It can then figure out what is the most power the panel can put out so it can charge the battery. Then it takes this charge and converts it for the best voltage, and you get the maximum AMPS to the battery.

Your MPPT controllers operate the most efficient, and you can use them with the higher voltage panels, thereby converting the unused portion of voltage to a higher amp. Some people have panels that put out a voltage that is higher than most of the standard RV solar panels. Because of this, they usually need to put in a quality MPPT so they can take advantage of the available higher voltage.

The MPPT will be the controller always to provide you the most yield. In the winter it could be as high as 20-30% more, while in the summer it could only be 10-20%.

The MPPT controllers can go up to 80 amps because they have the capacity, so they can handle extra amps that the PWM controllers cannot manage. Because it can manage additional amps, it allows more expansion throughout the system. Your main drawback is that the MPPT charge controllers cost a lot more than the PWM controllers.

There is also something called a 2-stage charger. It is a BAD thing. DO NOT let anyone talk you into buying a 2-stage charger. You will regret it forever. The problem being is that you can charge it up to a specific point and then without you touching it, you will see the reduction of power starting to come from the panels even though the batteries have not charged up yet. It seems to have the potential to suck the life out of your panels by taking back the power you have built up.

When you are deciding on the controller you want to spend your money on, think about the size of your system. If you have a small set-up, the difference in your efficiency between the MPPT and the PWM controllers may not be enough to justify spending that much money. But, if you are looking at a considerably more substantial system, your efficiency will play a more prominent role in how you want and need your whole system to perform.

As far as the MPPT charge controller goes let's look at it like this. The MPPT is a lot like the transmission of a car. If your transmission is not in the right gear, the wheels can't go the maximum power.

The reason is that the engine is either running too slow or too fast than what would be its best speed. The purpose of the transmission is to synchronize the wheels to the engine, this way the engine will run at a favorable rate despite the terrain changing and the acceleration.

To try and make this a little more straightforward, I am going to place the pros and cons of MPPT and PWM controllers on a grid form below.

Pros of the PWM	
	Pulse Width Modulator charge controllers have been used for years and are well established.
	Pulse Width Modulator charge controllers usually sell for under $250.
	You can find them in sizes up to 60 amps.
	Pulse Width Modulator controllers are considered durable, and most have passive heat sink style cooling.
	Pulse Width Modulator controllers are available in multiple sizes for all types of applications.
Cons of the PWM	
	The PV input minimal voltage has to match the battery bank minimal voltage if using the Pulse Width Modulator.
	As of yet, there has been no controller developed that is over 60 amps.
	You will find that there are many Pulse Width Modulator charge controllers have not been UL listed.
	There is smaller Pulse Width Modulator charge units that come without the fittings for conduit.
	Pulse Width Modulator charge controllers are limited in capacity for growth system-wide.

Pros of MPPT	
	Multiple Power Point Tracking controllers have the potential to increase the charging efficiency as much as 30%.
	Multiple Power Point Tracking controllers offer the possible ability to have the array with a higher input voltage than the battery bank
	Multiple Power Point Tracking controllers can be bought in sizes up to 80 amps.
	Multiple Power Point Tracking controller warranties are usually for a longer period than the PWM units.
	Multiple Power Point Tracking controllers have a better flexibility for system growth.
Cons of MPPT	
	Multiple Power Point Tracking controllers are expensive, and they cost two times the price of a Pulse Width Modulator controller.
	Multiple Power Point Tracking units are bigger in size physically.
	Sizing the right PV array can be very challenging without Multiple Power Point Tracking controller manufacturer specific guide.

Fusing Your System and Why it is Important

Let's take a minute and talk about how we should fuse the solar photovoltaic system.

It is necessary to know that the circuit breakers and the fuses primary use is to protect your system wiring from overheating and potentially catching on fire. They are used to protect your devices from getting damaged in the event of a short circuit and possibly keeping them from catching on fire.

Here is an example: You have a 12V battery that is lead acid. Let us say there is a short that develops in your AC/DC inverter; a fuse is located between the inverter and the battery that will prevent a potential explosion of the said battery and cut the circuit to prevent the wiring from getting dangerously hot or catching fire. The fuse saved the day for the battery, the wiring, and the AD/DC inverter.

Most commercially designed solar panels that rate over 50 watts will have a ten-gauge wire that can handle 30 amps of current. If you did connect these panels in a series, you would not increase the current flow, so fusing will not be needed for this string. If your panels are connected in parallel, your system current will add up.

If you have four panels and each of them is 15 amps, and you had a short in one panel, you can only draw all 60 amps towards the short-circuited panel. If this happens, it will cause the wires that run to that panel to exceed the 30 amps, and the potential is there for it to catch fire. If your panels are parallel, a 30-amp fuse would be required for each panel. If the panels you decide to use are smaller than 50-watts, you would only use 12 gauge wires and 20 amp fuses.

Chapter Five

Solar Panel Basics

If you are someone who has two cars, but the one is hardly ever driven, the battery in the car that is hardly ever started will eventually die. It is the same with your RV batteries except if you let them go uncharged for very long at a time it could cause you permanent battery damage and this would keep them from ever holding a charge again.

It's easy to avoid this situation by using a solar trickle charger. It will work at keeping the batteries at full charge and ready for you when you need them.

The trickle charger I refer to is most of the time a stand-alone unit that will create for you several watts for maintaining and charging your RV house batteries. Most trickle charges are made up of durable plastic, and the solar cells are amorphous. It is all molded into a unit that is weatherproof that measures almost a foot and a half square.

Installation of the trickle charger is straightforward and very simple.

It's a small panel, about the size of a dinner plate, you will connect the battery clips and wires and boom you are done.

Put your panel out in the sun, and you've begun generating electricity!

If you have a late model RV, more than likely one of these units will already be installed.

We should look at the difference between solar panels in parallel vs. series.

You probably are already aware the solar panels capacity in producing energy is measured in watts, and is then calculated by multiplying the voltage of the solar panel by the amps of the produced current. You must find the right balance of amps and voltage so that your system will perform at its very best.

If your panels are wired in a *series*, each of the panels will be connected to the others in a "string" like Christmas lights. By wiring in a series, you will sum the voltage of each solar panel and total them up, and the amps of electrical current will stay the same.

If you set up a series circuit, every panel will have to function for your circuit to be complete. If you have one panel go out in that series, it will break your circuit, and they will all go out. Your current will remain the same for each element in the series.

But, if you wire the panels in parallel, all of them will be joined to a central wire coming from the roof. Here, the amps of the electrical current for each panel are summed together, but your voltage for the whole system will remain the same.

I feel the only drawback is that there will be more wires running in from your set of solar panels. In this parallel circuit, each of your panels will have its own circuit, so even if all the panels go bad the last panel will still work. The total of the current for parallel setting is the sum of all the currents flowing through each component.

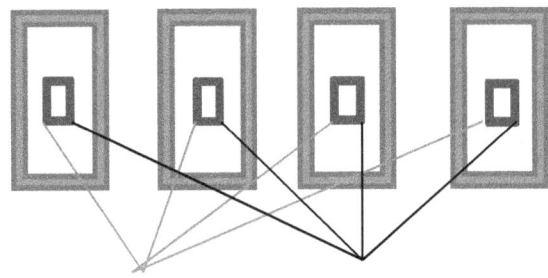

Now, for your **basic three types** of solar panels, you would use on an RV or van. They are:

- **Amorphous** – a thin layer of silicon that is attached to some type of backing material makes up each cell.

- **Polycrystalline** – this is silicon they have melted and poured out into a mold for each cell.

- **Monocrystalline** – a thin little wafer that is pure silicon crystal makes up each cell.

When we discussed trickle charges earlier; they were made up of amorphous cells. The amorphous type is the cheapest of all three, but also the best model to collect power for you on cloudy days. However, it isn't as efficient at gathering energy as your poly and mono cells.

You also need to know that most solar panels are designed specifically for charging and powering – as opposed to those used for maintaining charge. Those for preserving charge are comprised of either poly or mono cells.

Flexible or Rigid?

Whether your panels are made up of amorphous, polycrystalline or monocrystalline cells, the whole panel can be either flexible or rigid.

Most of the time you will see that the rigid frames are the made up of poly and mono cells. Flexible panels will be made from the amorphous, and even though they are less efficient, they can still be flexed up to 30 degrees.

Several companies on the market have developed new, more flexible panels that are made up of mono-cells that are probably worth considering; however, they are often quite a bit more expensive.

One interesting fact to be aware of is if your solar panel says it has a rating of a 12 V, you can bet it will probably be putting out 16-20 V. I know you must be wondering why the solar panels are putting out more voltage than what the batteries are supposed to hold.

If the panels had been designed only to put out 12 V, it wouldn't let them produce enough power during the less than ideal times and full sun. That is why the engineers design them to output that excess voltage than what is rated during perfect conditions. When you have bad weather or if your panels heat up (by the way your solar panels do work better when they are cool) you will still be able to depend on getting 12 V from them.

The solar charge controllers will read and then adjust as necessary the battery voltage, figure solar panel production, optimize power flow, and manage the RV's solar system and keep it running efficiently by making sure the batteries are not overcharging.

Your solar wiring is the mainstay of your entire RV solar power system. Don't let the sales people trick you into buying that expensive stuff. There is **NO** specialty wire to be purchased just for solar panels.

Mounting hardware should be done no matter what with the highest quality you can buy so that they stay secure in all situations. You have the option to place your panels on tilt brackets or lay them flat.

Some people think tilt brackets are a problem and prefer to place theirs flat. If you put your panels flat as a permanent fixture, you will be losing a lot of efficiency since you won't constantly be facing the sun.

Here I am talking to all you frugal, back to nature people and it makes me feel guilty. Don't get me wrong; I was taught to leave the world a better place than when I was born into it. But, with that said, I did go 21 days without any electricity due to an ice storm that only the movies can recreate about seven years ago, and I shall never forget it.

And, you know what I learned *really* quick. I LOVE comfort! I had roughed it while camping out on many occasions but never had I lived without electricity for 21 days in my entire life!

If I had my way I would have 100 solar panels set up all around my RV, but we all know that is not financially feasible or necessary. I am going to need panels that will give me enough solar power to save up for a week's worth of cloudy days.

How Tough Are My Solar Panels?

I worry about how tough my solar panels are when my RV is not under the shed built for it. What kind of nature's damage can they handle and what kind of damage can they take if or when I fall or step on them?

Here is what I found, and I was quite surprised since they are so thin. Manufacturers of solar panels routinely check their panels just for the situations that keep me awake at night. They launch chunks of ice the size of golf balls from high-pressured devices just to see what the panels can withstand.

It seems that at present most reliable manufacturers of solar panels that are certified can stand up to one-inch hail that falls at 50 mph. But, there are some that claim their panels aren't damaged with the same sized hail at 70 mph.

Where I live golf ball sized hail would be welcomed when at times we have had softball-sized hail. In areas where this happens, they urge you to consider using the thin-film (amorphous) photovoltaic panels instead any of the others.

The reasoning behind this is the substrate is a more flexible material made of plastic and is highly resistant to any damage brought on by rocks or large hail. According to the manufacturers, if any damage should happen that it would not have a measurable effect on your set of panels performance.

That sure is hard to believe, but they have proven it over and over in their testing facilities. I would just hope that it never happens because I can tell you how loud it is in an RV with hail and NO solar panels! If there were solar panels on my RV, all I would be thinking about is how I hoped the testers at the factory were right.

But, if there is damage, your solar equipment should all be covered under your homeowner's, RV, or van insurance policy. Make sure that you go over this issue with your agent and that has it included in your policy.

How about high winds? We see a lot of tornadoes here in the Midwest, usually followed by a good hailstorm, and then a lot of rain. We get to experience all of this on the same day! Your main threat to your solar panels at that time is the high winds and the driving rain.

But, again, the panels have all been tested in simulated situations to mimic hurricane force winds. Most of the solar boards can stand up to winds of 140 mph. You might find some loose panels and some loose conduits, but for the most part; the panels weathered the storm's winds.

What happens to your panels if you live in a region where you have a lot of snow and ice? Won't that be hard on the panels with all that extra weight? I live in the same place as the first two we talked about, and yes, we have snow storms that top 20" and we have had an ice storm of theatrical proportions that left 4" of ice in its wake.

What I found in my research was that solar panels are made to bear that weight! Their dark surface will collect sunlight thereby causing the melting of the snow or ice, and then it will run off the glass. When all the snow or ice is gone, the panel will go right back to gathering electric (energy) for you.

Sometimes people get frustrated with the flexible panels because it seems that when the temperatures reached 90 to 95 degrees that the thin covering was not good enough to keep the cells from "cupping." It made each cell form a shallow bowl that would collect dust in the bottom of the bowl, and it would block the sun from absorbing and cause a loss of power.

However, going flexible does have its downsides:

- You should never walk on these panels as they do crack under pressure as it will reduce their output of power.
- If you install flexible panels permanently then you give up their flexibility.
- No matter how delicate you handle them, they show signs of scuffing and scratching.
- The flexible panels only come with a ten-year warranty.
- Flexible panels do not have built-in ventilation to help the panels themselves stay cooler, so they get hotter and transmit all their heat right down into the inside of the roof they are mounted upon.

It seems the best way to place the thin but flexible panels is nothing but to glue them in place on your roof and when you get ready to remove them, it is not easy to get them off.

Chapter Six

Installing Your Panels

At this point, we are going to assume you have made smart choices on solar panel mounts, wiring, cabling, solar modules, inverters, charge controllers and everything else you will need. You have assembled every tool you have ever purchased, yes, even the old screwdriver you have been using to prop up that cupboard shelf within the kitchen. Isn't that where everyone keeps their screwdrivers?

I hope you have been reading this book so you know what we are about to do.

Let's talk about the location on the roof of your van or RV where you are going to put your solar panels. Be sure that it is in a place that will not be shaded anytime of the day. Do you have any raised TV antennas satellite dishes or raised roof vents? All of those can cause shading on your solar panels and kill your output of energy.

By now I guess you have picked out the perfect spot for your modules. The next step is looking for how you will route the cable inside the RV. One possibility is trying to go through the fridge vent. If you have a really large RV, some folks use the area of the holding tanks as their best choice. While you are deciding, do whatever you can to keep your wiring as short as possible so you will not encounter voltage drop.

Next spot you will need to pick out is the place for the charge controller. Some set them on the tongue of their trailer if it is a pull behind trailer. If you do place the battery on the trailer tongue, then you will probably have to mount your charge controller under the bed at the front of the camper on your front wall.

Since there is no need to look at it after you once have everything set up, it's fine to have it in a place that is not readily accessible. You want to put it as close to your batteries as you can but NOT in the same area as the batteries.

The reasoning behind not placing the charge controllers close to the batteries is because the battery gases over time can corrode and even destroy your electrical circuits on the solar charge controllers. You do not want to have to repair this part of your system in a couple of years.

Now, two 6 AWG wires would have to be run to this specific location from the solar panels from on top of the roof. Then two 4 AWG wires would have to be run from it down to the battery. Then you will need to run a ground wire from your charge controller to the frame. The mounting shunt that is inside and probably near the controller if you located it there would need more wiring.

Some folks mount the inverter next to their charge controller, and they get their power by tapping into one of the big wires from the battery running to the charge controller. They almost always put a 30A fuse inline on the hot wire to their inverter.

If it is a flush mount type controller, you will have to place it on an interior coach wall or at least a cabinet. If you are putting in a controller that has a remote display, there will be several more options. Some of them, depending on brand, can go in your underbelly near your batteries and then the remote display can be inside your RV.

Sometimes deciding where you mount the remote display is hard. It needs to be where you can see it quickly, but at the same time, you do not want the lights to bother you at night while you sleep.

Most of them are flush mount units, so you will need to leave about an inch of space behind it and enough of a path to place your wiring for it and run to the charge controller and your shunt. The display has almost a ½" overlap so it will cover any flaws you may have made while you were cutting your hole out.

This next step is determining where you want to put all your wiring. It can seem overwhelming but do not give up before we get started. I can assure that it seems that you can always find some little cavity or nook you had never noticed before to put the wire in from your solar panels to the charge controller.

It is the most frustrating and time-consuming part of the entire process. There will be some places you feel like you need a giant needle to run the wiring through a section to get it to the other side.

Now it is time to mount your solar panels. You need to park your RV or van out in an area where there is no shade for this part, and then after you decide where you want to mount your panels, run your wire.

If you are going to have flat fixed panels, you will want to run 'tray cable.' from the charge controller all the way up to your roof. Below is an example of a flat panel fixed kit.

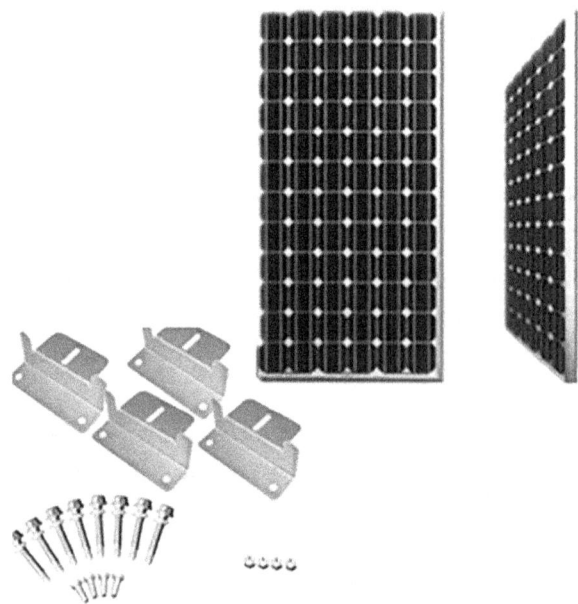

You are back on the roof again, and you will need to turn your solar panel over on its top on the roof of the RV and insert that tray cable thru your strain relief (waterproof) that you had already installed in your junction box. It is time to connect the red wire to the positive terminal and the black wire to the negative terminal.

Close the junction box, tighten the outer cap on your strain relief. If you happen to be placing more than three panels, it is wise to install a separate fuse in each junction box. If you are using flat solar panels, then install a fuse for every wire that is run whether in a dry location or a combiner box. See picture below of a strain relief.

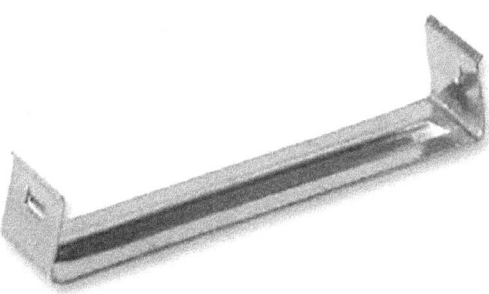

When you have completed this, turn your panels back over and attach them to the roof of your RV. It is essential that you take roof sealant and cover over the tops of the fasteners. For the next step, you will need to cover your solar panels with some of the cardboard from their shipping containers. You will need to tape, tie or clamp this cardboard to the solar panels, so you are blocking the sunlight. Fasten any loose wiring to the roof and beside the route to your charge controller.

Now it is time to install your charge controller. If it is a flush mount that goes inside the coach, make a cut out that is a tiny bit bigger than the template they included in the controller box.

If this controller needs battery power before the solar power, then that connection should be made first. The positive side of the wire (red) that runs from the controller to your battery will need to 'fused' at the battery. Now you need to make your connections for your solar input and uncover those solar panels.

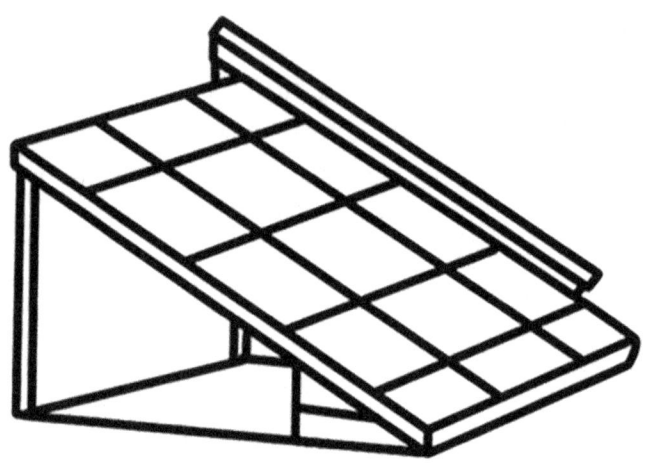

Tilting Your Panels

For those of you who put in the tiltable solar panels it is time we talk optimum tilt. If you want your solar panels to 'really' work for you, it is crucial that they are pointed in the right direction at all times of the year so that they will capture the most sun. It is not that easy, as there are variables that must be factored in when it comes to deciphering the best direction.

What I am about to try and explain will apply to any kind or brand of panel you might purchase that gets its energy from our sun. We will start by assuming that your panel is either fixed or has some type of tilt like the one in the diagram above, that you can adjust at different seasons.

(If you have the panels that track the movements of the sun all day long you will receive 40% more (summer) and 10% more (winter) in energy than with fixed panels. But as always panels such as this do come with a price.

Your solar panels should ALWAYS face the "true" south if you are in the northern hemisphere and the "true" north if you are in the southern hemisphere. Don't get confused at this point as true north is NOT magnetic north like we locate with our compasses.

If you are trying to determine true north with a compass DO NOT forget to factor in the difference that will vary from location to location. You will need to go on the internet and use the search terms *"magnetic declination"* so you can find what and how to apply for your correction in the location of where you will be staying.

Now, what angle from the horizontal position should we tilt the panels? Most advice offered, states the tilt should be at least equal to your latitude, minus 15 degrees in the summer, and plus 15 degrees in the winter.

The most straightforward way to place your solar units is a fixed tilt and just leave them that way. But with the sun higher in summer and lower in winter; it would be to your advantage energy wise to adjust your tilt with the season.

If you used a fixed tilt, which is convenient but has some disadvantages to it such as you will not be able to get as much power as you would if you adjusted the angle. If you happen to live where there is snow or ice, if your panels are adjusted, so they are at a steeper angle in winter it will help them shed snow better. If you have a panel with snow covering it, it will produce little or no power.

If you are planning on adjusting your panels twice a year, and you are working to grab the most energy over 365 days of the year, then read on.

If you live in the Northern Hemisphere adjust the summer angle of your panels on March 30th and the winter angle on September 12th of each year.

If you live in the Southern Hemisphere adjust the summer angle of your panels on September 29th and the winter angle on March 14th of each year.

Now if your latitude falls between _25 degrees and 50 degrees_, then you will find your best tilt angle _for the summer is your latitude x 0.93, minus 21 degrees_.

So, now, let us look at the tilt angle _for winter; take your latitude x 0.875, plus 19.2 degrees_. If your latitude falls outside the ranges available to you, it is time to go to the internet and research. The following link has a great solar calculator no matter where you live in the world at the moment to automatically figure your angle for you! http://solarelectricityhandbook.com/solar-angle-calculator.html

Maintenance of your Panels

Whatever you do, your panels MUST be maintained. By that I mean you must check them and make sure they are clean. If they are not clean, they cannot gather solar power for electricity efficiently. By cleaning your panels, it has been shown that you can even double your energy. For cleaning, you could buy expensive specialty soaps, but castile soap is an eco-friendly and cheap alternative.

Check with the manufacturer's directions and see if your panels have any special instructions for cleaning your panels. When you wash your panels, it needs to be on a cool day or very early in the day before it starts warming up and not while your panels are hot.

Before you start washing your panels, get the debris off of them first. Then use a soft scrub brush and soapy water and 'gently' scrub on your panels. I say be gentle because if you rub hard, you can cause micro scratches on the surface. Do not step on the panels while cleaning as this can cause micro-cracks that you do not need.

While washing or rinsing NEVER use a power washer; just a water hose with a sprayer will be sufficient for your solar panels. Then while your panels are still wet, use a squeegee and dry your panels.

After you have given your solar panels a good cleaning, you should notice a higher surge in the output of energy from your panels.

Chapter Seven

Size of Inverters
and How to Choose

*W*atts are a measure of the amount of power that a device uses when it is turned on. If you use a 100-watt bulb, that is just the *voltage* times *amps*. For example, of it draws ten amps at twelve volts or one amp at 120-volts, it will still be 120 watts.

Now, a watt-hour (or if you want to call it a kilowatt hour, kWh) is how many watts times how many hours that it is used. For example: If you have a light and it uses 100 watts, and you leave it on for nine hours, then that is 900 watt-hours.

An amp is another measure of electricity and its current at the moment. It does not come in "amps per hour.." Amps are essential as it is what determines the size of wire that you will need, mainly on the DC side of the inverter. Every wire has resistance, and as the amps flow through the wire, it makes heat up.

If you try to use a wire that is too small for the amps running through it, you will get hot wires, and that could run into a fire. It can also cause voltage drops if the wire is too small. An amp is defined by 1 Coulomb/second, and amp hours measure your battery capacity.

It brings us up to why we need an *inverter.* It must serve Surge or Peak Power loads, and usual or typical everyday power.

When there is a surge, that is the maximum of power the inverter can supply, but usually only for a short time. This short time is a few seconds up to maybe 15 minutes. There are some of the appliances, mainly those that have electric motors, which need a higher surge to start up than what they do when they are running. Prime examples are pumps and refrigerator compressors.

Typical power is what your inverter needs to supply on a continuous basis. It is a steady rating. Typical power is lower than your surge power. Average power is even less than your typical or surge power, and it doesn't factor in when you choose an inverter. Your inverter must be sized for the maximum peak load and for your typical load that is continuous.

Inverters come in several size ratings. From 50 watts up to 50,000 watts. When you look for your inverter, you will notice that they all have a surge rating and a continuous rating. Manufacturers usually specify it in so many watts and for so many seconds.

It means your inverter can handle an overload os the number of watts listed for a short period. The surge capacity will be different between inverters, and between different types, and then there will be differences even in the same brand. It may be hard to believe, but generally, a 3 – 15-second surge rating will cover almost 99% of every appliance.

You will find that the inverters that have the lowest surge ratings will be the high-speed switching electronic type and this is the most common. These are usually 25% to 50% in maximum overload. While high-frequency switching does allow a much smaller and a lighter unit, due to the smaller transformers that have been used it will also reduce the peak and surge capacity.

There are three different major types of inverters:

Sine Wave:

Sine Wave is the type of wave that you buy from your utility company or get from a generator. It's generated by AC machinery rotating, and sine waves are made from this rotation. The main advantage of the

Sine Wave is that all equipment you can go to the store and buy is produced for a sine wave. Unfortunately, you will also find that the Sine Wave inverter is very expensive – like 2 or 3 times more than the other types. Even though more expensive, it is the best for making your RV more like home.

Modified Sine Wave:

It has a waveform that is more like a square wave but has an extra step or two. The Modified Sine Wave will work with most of your equipment, except the efficiency and the power will probably be reduced. You will more than likely have problems with appliances like the refrigerator motors, fans, and pumps because they will pull more power from the inverter because of lower efficiency. Most of your motors will need and use about 20% more power. You will find that fluorescent lights will not be as bright and may make humming noises.

Square Wave:

There are not many of these, but they are the cheapest kind of inverters you can buy. They will only run elementary things, but not much else. You hardly ever see square wave inverters anymore.

How Much Total Power Will You Use in One Day?

For you to know what size system you are going to need, here are some much-needed formulas that you need to use to determine the total power you use in a day with your solar powered system.

- Watts = Amps x Volts
- Volt = Watts / Amps
- Amps = Watts / Volts

It is important to know how much current you have flowing to your load as it plays into selecting the wire you will need to buy. We must consider the distance to figure the voltage loss. You do not want to exceed 3% loss of voltage. Remember this; a larger wire is needed to move more current.

So, we have talked about a lot of things, but we still need to talk about the fuses and how to properly fuse your system. Fuses and Circuit breakers are there to protect your wiring in your system from getting overheated and stop it from catching on fire and well, whatever else is close to it. It also protects your devices from damage.

Here is a typical example: You have a 12V battery that is lead acid. There is a short that originates in the AC/DC inverter. The fuse between the inverter and the battery that will more than likely prevent an explosion of the battery and because of the fuse's location, it will cut the circuit quick enough to stop the wires from getting hot or catching fire. The fuse disabled the battery, AC/DC inverter, and wires and helped save all that damage.

For solar panel fusing, there are manufactured solar panels that are over 50 watts that have ten-gauge wires that can handle up to 30 amps of current flowing through them. If these panels are connected in a series, fusing will not be needed. If the panels are connected in parallel, and you have a short in one panel, it will draw all the amps to the short-circuited panel.

It will cause the wires going to that panel to exceed 30 amps and cause that wire-pair to catch fire. So, for panels in parallel, you will need a 30-amp fuse for each panel. Now, if your panels are smaller than 50 watts, and if you are using 12-gauge wires, then 20 amp fuses will be needed.

If you have the parallel system with a combiner box that holds your fuses and breakers for each panel, with one or even more "combined" fuses that lead to your charge controller or your grid tie inverter. When you size this 'combined' breaker/fuse, you will have to first decide on what could be worst-case of current that could flow based on your purchased panels.

If you have chosen a Pulse Width Modulated charge controller, your worst-case amps that are flowing back and forth from the controller are the same, so the fuse and the wire size can match in this case. MPPT charge controllers can increase the current that is flowing between the battery bank and lower the voltage, so the exact size wiring and fuse size will need to be recalculated, or you can find it in your charge controller manual.

A Note About Fuses and Wiring

Now, your fusing and wiring from the batteries to the AC/DC inverter is of the utmost importance as this is where the electricity will flow. Like the charge controller case, your recommended wire and the fusing should be looked up in the inverter manual to make sure you get the exact requirement. More than likely the invert will have a built-in fuse/breaker on the input section as well as on the output section side of the unit.

The typical 1500-watt 12V pure sine wave inverter is found to draw up to 125 amps continuously, for a number that increases to 156 amps once we look and factor in the NEC continued to use 25% adder. For USE-2 wires, 1/0 AWG would be required in this circumstance.

    #4 AWG 150 amps

    #2 AWG 200 amps

    #1 AWG 250 amps

    1/0 AWG 300 amps

    2/0 AWG 400 amps

Factors like cable length, and fuse and breaker types need to be examined before beginning this process. Luckily, there are numerous resources online that deal with wire size and fuse size calculators.

Let's take a look a quick look at what we have talked about so far, starting at the top of our RV or van and going to the inside finish and ready to go Solar Unit.

- We have our Solar Panels (usually on top) that will take the sun's rays and convert it electricity that will run on down and charge up the batteries blow. These will be fitted with your mounting brackets and cables.

- Then we will have our cables running through conduit running from our solar panels over to the entry plate for our cable that will help this area be vibration proof and watertight.

- The wires will travel on down to our installed solar controller that is used to regulate how much electricity is brought in from the panels and sent into the batteries. It keeps the batteries from overcharging.

- Most of you should also have a battery converter/charger, so if needed you can hook up to shore power for charging your batteries if necessary.

- There will be a transfer switch next to the battery converter that will help by 'automatically' selecting *shore* or the *inverter power* to give to the AC breaker panel.

- Next, there will be an inverter (I did not say 'converter' again) that converts DC (12V) which is battery power over to AC (120V) power that can be used to power household appliances, electronics, and devices.

- You will notice there is an inverter fuse with a cable kit nearby.

- And finally, there will be your battery bank. It will consist of your high-efficiency batteries (anywhere from 1-4+ will be needed) and are the power reservoir for your system. This bank of batteries will store your electric supply from the solar panels.

For me, I would love twelve batteries so that I could have lots of energy stored up. No, I have no idea where I would put them, but maybe someone could weld me a special box underneath to fit them all in.

Yes, they would be hard to get to, but I would have energy when I wanted it. Let me add that you do not have to buy all these parts piece by piece as there are many ready to install kits to set up your mobile unit and go!

Chapter Eight

Solar Panels and Lightning
Are Not Friends

Unfortunately, many have found out too late that lightning is one of the most common causes of failure with photovoltaic systems. A surge can hit from lightning that will strike a long way from where the system is even located, or between the clouds. But, you can prevent most lightning damage.

The following will give you some of the most cost-effective ways that will work for your system based on years of experience. Just follow this advice, and you should be in good shape of avoiding damage from lightning to your solar energy system.

Grounding is the first and foremost technique to protect your system from lightning damage. You can give lightning a new direct path to bypass all your valuable equipment and discharge safely into the ground.

Lightning arrestors and the famous surge protectors we find in all homes are designed to protect our expensive electronic equipment by absorbing those electrical surges. These surge protectors are not a good substitute for grounding.

The surge protectors function only in combination with suitable grounding, while the grounding system itself is an essential part of the wiring foundation. You should install it while or before the power wiring is connected. If you don't do it then, once you get your system working, you may forget to take care of this essential component.

The first thing you must do in grounding is to create a path of discharge to the ground by interconnecting all the electrical enclosures and metal components, like the PV module frames, the wind generator towers, and the mounting racks.

You can check the National Electrical Code in Article 250 and again in Article 690.41 thru 690.47 to find out what the code-compliant wire sizes, techniques, and materials are that are required. You must avoid sharp bends with your ground wires as high current surges do not like turning tight corners and will quickly jump over to nearby wiring.

Pay extra particular attention to the attachments of copper wire to any aluminum structural elements. Use connectors that are labeled "AL/CU" and stainless-steel fasteners that will reduce the chance of corrosion. The ground wires for both AC and DC circuits will be connected to this grounding system.

The weakest link of many installations is connecting to the earth itself. It is necessary to bury or hammer a rod of noncorrosive metal (usually copper), that is conducive to the ground, and then make sure most of its surface area contacts moist soil. So, when there is static electricity, or a big surge comes down the line, the electrons will drain into the ground with little or no resistance.

Install one or even more 8-foot-long, and at least 5/8-inch copper-plated ground rods into the moist earth. A single rod is usually not enough. If your ground is arid, install a lot of rods, spacing them at least six feet apart and then connecting them with bare copper wire that is buried. Now, there is an alternate approach that can be taken, and that is to bury #6, double #8, or larger bare copper wire in a trench at least 100 feet long.

If you can, try to route the system into wetter areas, like where plants are watered all the time or where the roof drains.

In moist climates, you will find that concrete footers of ground- or pole-mounted array, a wind generator tower, or ground rods encased in concrete will not be adequate for grounding. If this is you, then you should install a ground rod in the earth next to the concrete at the base of an array, then connect all of them with bare, buried wire. Just remember – you CANNOT have too much grounding.

Chapter Nine

Wiring Your RV – Mobile Unit

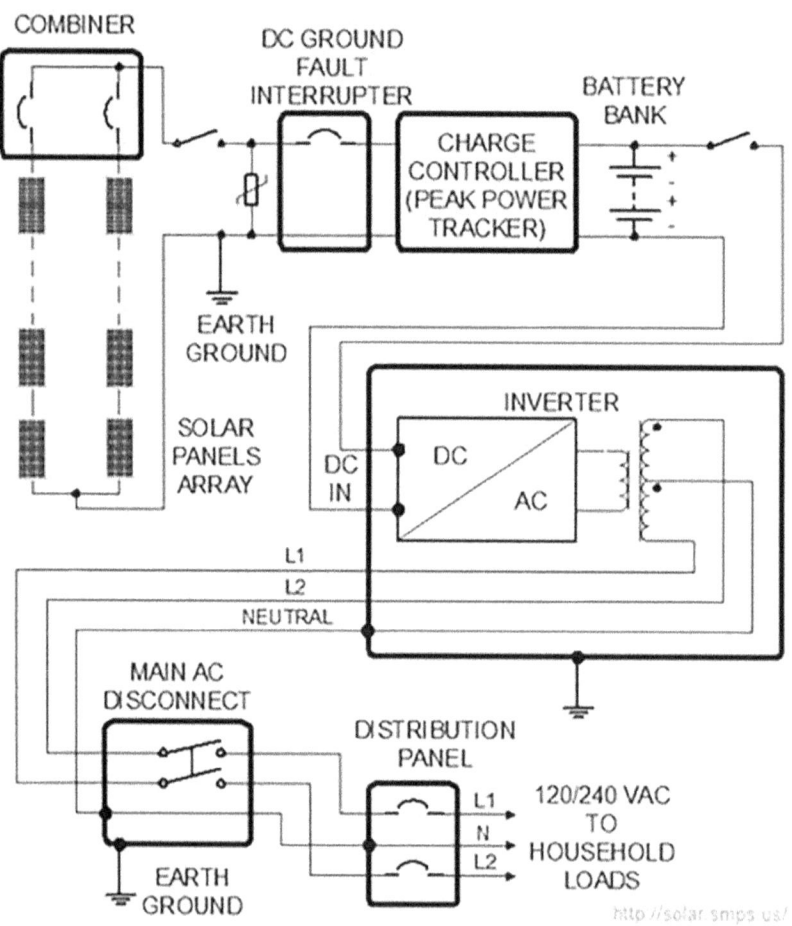

Here is a wiring diagram of Mobile Home Solar Panel System. It is a fundamental diagram that I feel will give you the general idea of how to wire your RV or Van. Of course, your panels would be a layover at the top of the diagram in real life. If you happen to choose a kit, it may come with a similar diagram like the one above for you to look at as a reference.

Here are some standard instructions below on installation of a mobile unit with three panels located on the top of your RV or Van.

- Take off all your jewelry until this is done.
- Disconnect everything from its source of power.
- Cover your solar panels with a quilt or heavy blanket until you are finished with the installation.
  - Layout your panels and your wire like you want them to be but make sure:
    - That the three-port roof cap is within 15 feet from your charge controller
    - Your panel leads can reach the three-port roof cap
    - Panels are not being shaded by AC units, vents, any other obstructions
    - Charge controller is not any more than 9" from your batteries
- Take the 15' solar wire and connect it to the three-port roof cap. (The red wire goes on the silver terminal, black wire will go on the black terminal, but the red terminal will not be used.)
- Start feeding the wires through the roof and mount 3-port roof cap. Finish routing your 15' cable from the roof to the charge controller. DO NOT make connections at this point.
- Now, route the 10' cable from your charge controller to your battery. Again, DO NOT make connections.
- Connect your + solar wire to your charge controller, connect the + battery wire to your controller. Do the same for your negative wires.
- Mount your charge controller.
- It is time to attach the mounting feet to the panels and proceed with the panel install.

- Connect the + wire to your battery and do the same thing with your negative wire.

- You can now plug your SAE leads into that three-port cap, and now you can uncover your panels.

- Your panels are now charging your battery!

- Configure the charge controller, so it is set to the right battery type.

The following page will give you a couple of diagrams that show you wiring paths that were used for the two RV's that are represented.

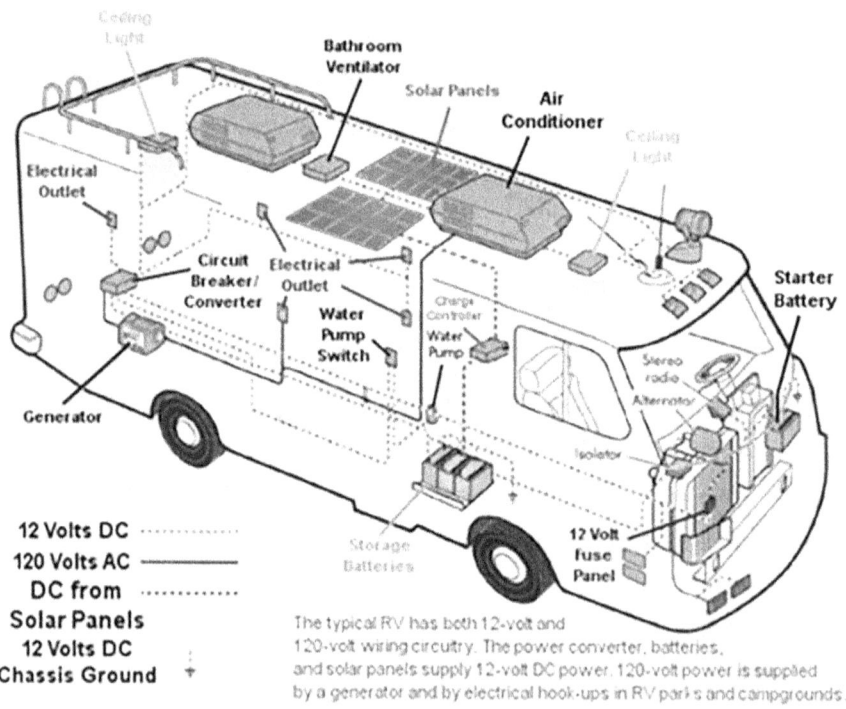

Ceiling Light

Bathroom Ventilator

Solar Panels

Air Conditioner

Ceiling Light

Electrical Outlet

Circuit Breaker/ Converter

Electrical Outlet

Charge Controller

Water Pump Switch

Water Pump

Starter Battery

Stereo radio

Alternator

Generator

Isolator

12 Volts DC ············
120 Volts AC ──────
DC from ············
Solar Panels
12 Volts DC
Chassis Ground ⊤

Storage Batteries

12 Volt fuse Panel

The typical RV has both 12-volt and 120-volt wiring circuitry. The power converter, batteries, and solar panels supply 12-volt DC power. 120-volt power is supplied by a generator and by electrical hook-ups in RV parks and campgrounds.

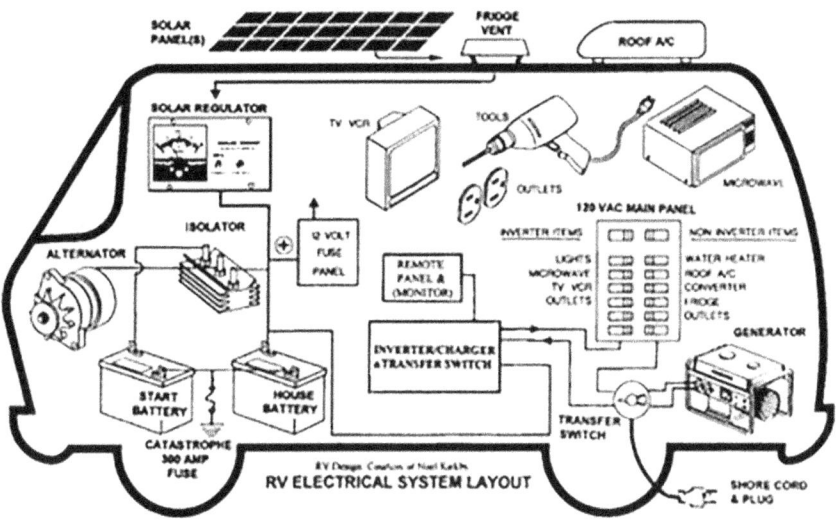

RV ELECTRICAL SYSTEM LAYOUT

Final Words

This book has been enjoyable to write and share with others on how to live off the grid with comfort. I admire everyone who is willing to make the world a better place and live 'off the grid.' I salute you.

I also know when you finish this project that you will be proud of what you have accomplished. There are many couples who have completed this task together by simply following the directions. They are very happy with what they have done together as a team.

You will meet some of these couples as you hit the road and you will find that they are more than ready to share their stories with you about their experiences of solar panel living on the go. Most will tell you it can't get any better than this and that they wouldn't have it any other way. My hat is off to them!

They have been chasing their dreams and making them a reality as they should. I think it is extremely important that everyone live life 'in the moment' and have the opportunity to live life in a large way even if it takes being frugal. Doing both is possible.

Never let a chance pass you by and wish later that you had passed on it. You will always have regrets. Live your dreams and don't give up.

The only thing in this instance is, you are a step ahead of those that have *not* gone before you. You have read a book that will save you weeks and months of researching and testing in preparation for your project.

Good luck my friends. Enjoy your time as you prepare for your escape to enjoy nature and to watch a different night sky whenever you feel like it. I wish you nothing but the best.

I want to thank you for reading my book and if you enjoyed reading my book could you please do me a favor and leave a review?